An Orderly Mess

Helga Nowotny

An Orderly Mess

CEU PRESS

Central European University Press

Budapest–New York

Published in 2017 by
Central European University Press

Nádor utca 11, H-1051 Budapest, Hungary
Tel: +36-1-327-3138 or 327-3000
Fax: +36-1-327-3183
E-mail: ceupress@press.ceu.edu
Website: www.ceupress.com

224 West 57th Street, New York NY 10019, USA
E-mail: meszarosa@press.ceu.edu

ISBN 978-963-386-231-5

Library of Congress Cataloging-in-Publication Data

Names: Nowotny, Helga, author.
Title: An orderly mess / Helga Nowotny.
Description: New York : Central European University Press, 2017. | Includes bibliographical references.
Identifiers: LCCN 2017040740 | ISBN 9789633862315 (pbk.)
Subjects: LCSH: Order (Philosophy) | Uncertainty. | Forecasting. | Future, The.
Classification: LCC B105.O7 N69 2017 | DDC 117--dc23 LC record available at https://lccn.loc.gov/2017040740

Printed in Hungary

Table of Contents

Foreword and Acknowledgements 1

An Orderly Mess 5

Eigenzeit. Revisited 61

Foreword and acknowledgements

Trying to make sense of the unexpected developments that have recently shaken some of the complacent assumptions of the Western world made me think about messiness. Like many others, I puzzled over questions such as "Would we have predicted that...? Could we have known? And if we knew, why did we not act?" while facing the contradiction that the "we" in these questions has itself become highly questionable. Some of us are beginning to realize that democracies too can die. The Cunning of Uncertainty* is needed as an ally more than ever before.

Given my background in science and technology studies (STS), I wanted to situate the contemporary phenomena that I came to call "messiness" in a larger frame.

In physics, the term "phase transition" is used to describe transitions between different states of order. They involve symmetry-breaking, first- and second-order transitions, critical points, and in most cases, abrupt change. In all these the process of going from one state of order to another culminates through a maximum disorder. Messiness is needed to give way to a transition that would otherwise not take place. The more a transition is needed to reach a new order, the more messy is the transition.

Sydney Brenner, Nobel laureate in physiology and medicine, once remarked that mathematics deals with the perfect, physics with the optimal, and biology with the satisfactory. This leaves out the social sciences. They deal with the messy.

I was able to finish my attempt to work out not only what produces and defines messiness but where it leads during my stay as visiting professor at Nanyang Technological University in Singapore. In particular, I want to thank Alan Chan, dean of the College of Humanities, Arts, and Social Sciences, for raising the question of whether there can be an optimal balance of messiness. The ideal is appealing, although—like living with what evolution has bequeathed to us—we might have to settle for what is merely satisfactory.

The second essay, "Eigenzeit. Revisited," was originally published in German.** It is based on the opening lecture Bernd Scherer invited me to give at the inauguration of the project "Hundert Jahre Gegenwart" at the Haus der Kulturen in Berlin in September 2015. Liviu Matei, provost of Central European University, and Krisztina Kós generously invited me to publish the English translation in an expanded version with CEU Press. My dear friend Giovanni Frazzetto agreed again to act as a sounding board and suggested the title.

My warmest thanks go to all of them.

Helga Nowotny
Singapore, February 2017

* Helga Nowotny, *The Cunning of Uncertainty* (Cambridge: Polity, 2015).

** Helga Nowotny, "Eigenzeit. Revisited," in *Die Zeit der Algorithmen* (100 Jahre Gegenwart), ed. Bernd Scherer (Berlin: Matthes & Seitz, 2016).

An Orderly Mess

Why we need messiness

Messiness is a deeply ambiguous concept. Hovering somewhere between order and disorder, it defies precise description. Among those who experience it, tolerance of messiness is bound to vary greatly depending on the situation.

Messiness is utterly familar to us. It surrounds us in the intimate space of the home, be it the kitchen, the desk, or the bedroom. It pervades our oscillating emotional relationships to parents, siblings, and Others as we grow up. Messiness functions as background noise. Most of the time it is taken for granted, but it may become a major irritant once the delicate balance is upset between the forces that disturb order and those that maintain it. Then we realize that messiness is yet another manifestation of our always fragile relationship to a world that extends beyond our control. It then becomes the testing ground for our resilience to withstand disorder while unleashing the creativity in doing so.

Messiness is hardly new yet surprises us every time. It mirrors a state of transition that is as temporary as it is fundamental. For messiness is akin to a state of limbo: it is unclear whether it will tilt towards order or disorder. Unbending to a sense of time, its conceptual as well as its experiential boundaries are fuzzy and ever-changing. Yet given its fluctuating and restless nature, it is impossible to escape.

In fact, there is also a pleasant side to messiness. It can unleash creativity. Alison Gopnik argues that "from brains to babies to robots to scientists, mess has merits. A system that shifts and varies, even randomly, can adapt to a changing world in a more intelligent and flexible way."[1] Messiness can demonstrably boost the ability to work in teams. It has a special charm for those who abhor too much tidiness, as it sets free bursts of creative energy. As such, it offers a taste of freedom from social norms and societal constraints. It can act as a subversive force that undermines pedantry. The playfulness it conjures up can get people together to work on ideas and induce action that any single member of the group would have been unable to achieve.[2] It is a gateway to the creative exploration of a world in which we always look for—and find—patterns even if we encounter only randomness.

But there are downsides as well. We may enjoy a day-to-day messiness as long as it does not undermine the stability of social expectations and structures that we take for granted. Once the outer surface bursts and the taken-for-granted neatness dissolves, messiness fully reveals its volatile side. It begins to slide, or wobble, towards a darker, disorderly state where things can easily get out of control. The trigger for this often remains hidden. It may be a slippery slope, but one that does not follow a linear path. It may be a sudden and unexpected event that causes dislocation and seeds confusion. In the end, chaos is waiting. It engulfs indiscriminately everything in its path.

In these times of political turmoil, many of us feel steeped in a particularly nasty and widespread state of messiness. But civilization, at least as we know it, has not yet broken down. Despite "alternative facts" and the wild, unrestrained circulation of fake news, most of us still hold on to factual knowledge as our most trusted anchoring of reality. Bewildering as it sometimes seems, we believe that reality is fully independent of our will and yet open to our intervention. Messiness lies precisely in between the two. Yet we experience its dynamics as unpredictable. The playfulness may suddenly turn into a threat. Paraphrasing the lyrics of Bob Dylan, we may ask ourselves: if I accept messiness, will messiness accept me? In other words: can I trust messiness? Instinctively, we crave some kind of assurance that we will emerge from troubled waters again.

For societies this proves to be more difficult. And there are no assurances, no guarantees. With xenophobia, nationalism, and populism on the rise again in many countries, and reason trampled upon and overtaken by anger and emotion, it is as if ghosts from the 1920s and 1930s have come to haunt us. These ghosts of the past re-emerge along with geopolitical tensions that remind us of the horrendous developments that followed that era. Upon seeing the ghost of his father, Hamlet reacts by exclaiming: "Time is out of joint." His later curse, "O cursèd spite/That ever I was born to set it right!" now falls upon us. How will we in the twenty-first century defend ourselves against these ghosts of the past?

The cosmopolitan and liberal elites stand accused of having become disconnected from "the people," while another part of the wealthy establishment looks on with glee. Populists proclaim that they, and only they, know who "the people" are and what they want. Experts have been denigrated and declared obsolete. This tumult in the political arena recalls a medieval Carnival ritual. On "fool's day," servants acted as masters, and men and women reversed their traditional roles. But while this folly was limited to one day, we seem to be in it for longer, with no end in sight.

The joyful innocence of messiness is liable to turn towards a disordered ending. In the rupture of confidence and trust in public institutions and the elites that run them, science and scientists are implicated as well. Undoubtedly, they are part of the elites. They are international in outlook and connections and hence suspected of limited loyalty to their home country. They too—as so-called experts—are an easy target for populists who brush them aside as arrogant and unempathic. Together with expertise, scientific facts are in for a difficult time. They can be dismissed as a mere sub-category of other facts and their makers accused of either acting on behalf of interest groups or simply being naive and over-cautious. The reaction the scientific community has developed over time when it meets disbelief or feels under attack is to call for more and better communication. Yet this old Enlightenment defense has turned into a rather blunt weapon. Science and democracy have not always had an easy and tension-free relationship in the past. But now, perhaps for the first time since the end of World War II, both seem endangered. As the fundamentals of democracy are under threat in the United States, operating as a free, democratic, and powerful nation, so are the fundamentals of science and how it functions. Messiness might soon turn into a big mess.

We are rapidly discovering that it will not suffice to counter alleged facts with real facts. The dividing line is quickly eroding between fact sheets and cheat sheets. "Trust in numbers" (in Roy Porter's phrase), which was one of the hallmarks of the modern nation-state, is rapidly collapsing, at least in the sense that statistical data are losing the authority they once held. This occurs on a double front:

quantitative data as a basis of factual evidence are distrusted by populists and by people who distrust their government. So are experts who use them to legitimize policies, evidence-based or not. The second front on which statistical data stand to lose is the one opened by predictive analytics and the big-data pool from which they are extracted. Statistics are collected on the basis of knowing which questions to ask. The categories to be used are determined in advance, usually by governments and from above. The data on which predictive analytics are based have a different origin. They result from the sweeping collection of all the traces we leave behind in a digital economy and society. Big data have a built-in flexibility that allows them to be used to answer questions they were not collected for. They can be repurposed.

Caught somewhere in between, quantitative evidence has lost much of its clout. Above all, it has lost the authority and legitimacy that had been provided by the state as the ultimate guarantor for its factual veracity. This, at least, was the ideal. Instead of serving a predefined purpose to inform the state administration, the answers Big data provide are primarily for the profit of the corporations that own them but also for the individuals whom they target. Instead of "being counted" as a passive unit, consumers are encouraged to believe that they themselves produce the data that yield the answers and the products they are seeking. The illusion consists in believing that users have escaped all administrative and demographic preordained categories. They feel they can now generate their own collectivities, in a bottom-up mode, by selecting which networks and social groups they want to belong to just by pressing a "Like" button.

But the sense of freedom from bureaucratic control is illusionary. The question is not about the loss of trust but about whom and what we trust. Statistics have not vanished. Numbers are reasserting themselves as indicators covering a wide range of activities. They act like boulders, stemming a tide of messiness when the old boundaries between state and market, between private and public, no longer hold. The "new public management" approach, originating in the private sector, has been transplanted to make the public sector more efficient. Numerous key performance indicators (KPIs) and other benchmarking and best-practice measures are in place in order to control performance and ensure its continuous improvement.

But messiness enters through the back door. Indicators are built on present goals, with predetermined points in time for assessment and evaluation. They disregard what went on before and afterwards. They ignore the processes and activities that slip through the tightly woven KPI net. These processes will surface elsewhere, sometimes with eruptive force. Then they will override the KPIs, who will appear outdated, if not obsolete. So even if the surface seems smooth and under control, messiness lurks below. It reminds us of the interplay honed over time between the visible and invisible, the formal and the informal. Messiness asserts itself by refusing to take sides and thus provides the necessary link between the two sides.

With the public space decreasing and being hollowed out, the private space is left for compartmentalization to take over. Bubbles form that are self-closing social environments. This is greatly enhanced by the ease of communication through the ubiquitously accessible social media. Paradoxically, the wider social networks spread, suggesting unprecedented openness, the more users want to be protected from any views that clash with their own. Due to confirmation bias, one prefers information that fits with what one already knows and believes.

Communication in self-closed and self-restricted spaces is deprived of the ambiguities that would otherwise allow *Anschlussfähigkeit,* the capacity to connect and establish a fit with persons or objects, opinions, or beliefs that may at first seem incompatible. These are attempts to exclude messiness. At the extreme, a kind of cleansing mentality takes over, with all the nefarious associations that the concept of “purity” has contained in the past. This bodes ill for the future. For what begins with the marginalization of the multiplicity of meanings may open the door to eliminating the multiplicity of beings.

In the following sections I want to situate examples of contemporary experienced messiness in a longer transition. It is not a linear process, even if one cannot escape the linearity imposed by the narrative. I will explore messiness at work in the two major sites that coordinate social life: time and space. They are under stress and yielding. Unpredictability is the name of the game. The key to understanding the transition are novel attempts at world ordering. These are ways through which societies in earlier times and in a variety of places attempted to cope with the messiness they encountered. They are attempts, often futile and sometimes successful, to keep disorder at bay. They are deeply steeped in practices. Modernity, deeply interwoven with modern science and technology, created powerful processes of world ordering that are conceptual as well as instrumental and political, eventually conquering the globe.

This modern way of world ordering is in the process of undoing itself. One of the most powerful dichotomies on which the vast empire of modernity was built was the separation between human and artificial and between the natural and human-made. The artificial is now invading and encroaching upon what used to be the domain of the natural and human.

This is largely due to the advances in science and technology, but they cannot be isolated from society. Technologies change us through the use we make of them as much as we invent and change them. Science is a unique cultural activity, strongly aligned with the societies that cultivate it. As with modernity's world-ordering, science and technology may instigate powerful changes but do not control them. In the closely entangled processes of co-production and co-evolution that link science and society, computers and computational thinking change the way we live, work, and relate to each other, as well as to the natural environment. The contours of a novel way of world ordering are barely visible. What can be seen is that it is more open to accommodate messiness—not as an unwelcome intrusion or threat but as an integral part of how things are.

We need messiness to remind us that our existence remains poised between order and disorder. It may, at least ideally, protect us against too much and too little of both. It holds out the promise of an always precarious, but perhaps achievable, optimal balance between them.

Broken timelines

Messiness is located in time and space, in a physical but also deeply social dimension. Nothing in human existence escapes time, and whatever we do is spatially bounded. The biological arrow of time accompanies us from birth to death, as we struggle to come to terms with the finitude of our existence. Elaborate conceptual temporal schemes invented by indigenous people around the globe served to coordinate their environmental survival niche with their cosmological and social order. Historians have explored the distinct temporalities that prevailed in different historical periods. The concept of an open future arose in Europe only in the late 18th century, when expectation of what was yet to come began to differ from past experience. The age of industrialization ushered in the dominance of clock time and of linear temporal sequences. The old dream of simultaneity seemed finally to have come true with the latest wave of globalization.

This temporal order is now dissolving. This is partly due to the neoliberal drive for greater efficiency and performance management. It is also the result of growing complexity when different elements of a system with their own temporalities interconnect and often clash. Once again, the recurrent feeling, from Hamlet, that "time is out of joint" is not unique to the present. It arises whenever expectations are invalidated and when unexpected events seem to tumble out of nowhere. Then the moment arrives when nothing seems to hold together anymore, because reliance on a habitual “before” and “after” has gone. The comfortable reassurance that expectations will be honored has likewise vanished.

Today we live with broken timelines. While we have learned to accommodate some, we feel a deep unease or even pain over others. Nowhere is this more evident than with the timelines that used to structure work and hold work and life together. Long gone are the days when those who had work could expect to keep it until retirement. The eight-to-five working day and a working schedule on fixed days of the week have become the exception. For the younger generation, the future of their working life looks precarious, and for those with low skills and education, it is bleak. With the automation of middle-class jobs rapidly proceeding, adjustments in terms of blending the temporal patterns of human-machine work mean even more cracks in the timelines.

Then there are the broken timelines of the past. They persist as the scars and memories from when some major time strands holding together Western civilization were torn apart. The first big shock came with World War I and continued with the horrors that followed, leaving their indelible mark on the 20th century. These profound ruptures brought an end to the belief in progress and to the grand narratives invented to support it. The linearity of history became questionable, as modernities around the world started to multiply on their own terms instead of following one model, the dominant Western one. Multiple modernities with different temporal trajectories inscribed in them and with different outlooks on the future are now shaping the geopolitical landscape, driving and driven by technological and economic globalization.

Meanwhile, humanity and the dream of global citizens are confronted with long-term timelines on the verge of breaking: the rapid degradation of the natural environment and the benefits derived from it. Climate change and the already observable disastrous fallout confronts the entire world with a temporal complexity at an unprecedented scale. Symbolized in the Doomsday Clock, with its warning of how close the world stands to "midnight" (a nuclear or other existential catastrophe), the sense of urgency is clearly felt. Set up in 1947 by the Bulletin of Atomic Scientists and regularly readjusted since, it has just been reset following the election of President Trump to two-and-a-half minutes before midnight—the end of all human timelines.

The "cursèd spite" to have been "born to set it right" is also evident in the discussion about a new epoch, the Anthropocene. Geologists are the official scientific timekeepers of the timelines of the earth, the geological epochs that made and change it. From the late 18th century onwards, they were at the forefront of the effort to uncover the earth's various strata and to determine its age. But there was more to it. The strata contained much-sought-after minerals. They were home to the remnants of the organic waste that had been transformed into carbon. Geologists explored the earth's inner layers in order to extract and transform this material into the energy that would fuel the Industrial Age.

In the twenty-first century, geologists are again at the forefront of time-setting. A working group of the International Union of Geological Sciences has been set up to determine whether sufficient stratigraphic evidence exists to warrant the demarcation of a new epoch, the Anthropocene—the age when humans began to leave indelible traces on the vast domain of nature. Some argue that the radioactive traces left in rocks after the first trial explosions of atomic weapons constitute such evidence. Others want to set the starting date at a much earlier stage, when humans began to ship plants and animals across the oceans, altering species habitats in irreversible ways that culminate in the rapid loss of biodiversity we witness today.

Whatever the ultimate outcome will be, the beginning of a new epoch must be acknowledged in scientific terms. While previous beginnings and endings of epochs were brought about largely by natural catastrophes and large-scale discontinuities, the decisive and long-lasting impact of human intervention in the natural environment marks the breaking of a major timeline by humans. From now on we will have to contend with the complex temporal processes and pattern in oceanic and atmospheric circulation, the time scales of radioactive waste degradation, the rate of acidification of the oceans, and many other interconnected processes—all directly and indirectly linked to the unintended consequences of human action.

This leads to the temporal dynamics of complex adaptive systems. Complexity arises from the interlinkages between the component parts of a system. Entire new phenomena may emerge, but entire systems may also collapse. They may be ecological systems or financial systems, electric grids or systems of urban transportation, a health system in crisis mode, and even failed states. The temporal dynamics of a complex system has tipping points built into it. Entire previous civilizations, including vastly extended agriculture-based, low-density urban empires like the Great Angkor region and the Maya civilization, ended in collapse. As the critical mass of data points accumulates, one last triggering event may lead to an abrupt change. Human cognitive capabilities to detect sudden changes in advance, let alone to act and prevent them, are underdeveloped. These are the time zones to be explored by simulation models and the simulated worlds of artificial agents that are made to act on our behalf, allowing us to see a bit further. Yet the present understanding of the times of complexity, of the intersection between short and long processes and their distinctive temporal regimes, is still limited.

The working of the human brain also follows some peculiar cognitive biases. Humans are known to discount the future by valuing present assets much higher. Politicians are known to adjust their future horizon in line with election cycles. Business is known to shrink from the foreseeable threats that come with climate change. All of us extend our own timelines into a future that is fraught with uncertainties about who will be most affected, where, and how. But knowledge does not automatically lead to action. There is a tenuous, often contingent link between the two, because knowledge must become contextualized if it is to be translated into action. Knowledge is always knowledge for what. It must be interpreted to suit the practices, and practices must be ready to integrate knowledge.

Human temporal anchoring occupies the middle ground of time scales carved out by biological evolution and enabling human existence. The time generators of societies reside in institutions. Their temporal regimes coordinate, regulate, and impose the temporal grids and timelines that structure our lives. The *eigenzeit* of individuals often clashes with the temporal demands emanating from institutions. Conflicts arise when different temporal regimes, articulating the demands and control mechanisms institutions deploy, clash among themselves. The individuals caught in between experience such confrontations as cracks in the timelines, eventually leading towards a temporal breaking point.

Academic life provides ample illustrations. The "theaters of accountability," as Ulrike Felt calls them, continue to multiply. The pressure to publish in the most highly cited journals with the highest impact has become enormous. In the name of transparency, better self-control, efficiency, and the manageability of the research system, new temporal regimes have been introduced into academic research. In particular, the growing number of young researchers on short-term contracts are caught in a tight temporal grid. The new metric imagination tied to article production seems to offer a quick validation of their work and access to funding for the next project, while fitting with the temporal and counting logic of funding agencies and universities.[3]

This leaves no time for maturation.[4] It ignores the fact that finishing an experiment may not coincide with a deadline for publication. The life cycle of the model organism in the lab may not fit with the date set for the promotion committee to meet. The writing of a book follows its own temporal growth pattern, with punctuations between fast and slow, bursts of ideas and slack. Especially for younger researchers, this disjuncture of temporal regimes means being caught between incompatible demands. To pursue their career, they must either adapt or get out. Time that is relatively free from institutional demands has become a scarcer resource than ever. Timelines in academic life have become brittle.

This is not the whole story, familiar as it might seem. The world consists not only of academics but also of many others who are pressed into irregular but predefined time slots. They are expected to deliver within disconnected chunks of time, set to be exchangeable and fungible while devoid of content. These timelines are reminiscent of the old industrial manufacturing piecework, only that much of the actual production line is now automated or implies working with software that controls the automated processes. Growing numbers of people are left to fend for themselves. Their time lines are precarious, but precariousness has turned into a steady state. They have to find their next job, which is limited in duration, as was the last one.

Meanwhile, there are age limits and limitless aging; there is time spent on social media and the visions of an imagined future just around the corner, when commuters will be able to "work on the go," sitting in driverless cars that will ferry them around. Time use driven by efficiency and profit maximization slides, seemingly effortlessly, into time used for scrambling frantically to keep things together.

Debates on determinism and free will, on the block universe—a cosmological theory according to which time only appears to pass—and on the reality of time, have lost much of the fascination they once held for a larger public. They have been replaced by heated public discussions on the potential threats emanating from artificial intelligence (AI) and its capacity to take control, on whether and when such a point of singularity will be reached, and what an engineering approach converging with an organic approach of life will mean. Advances in the neurosciences have curbed the confidence in free will, and new genetic tools like CRISPR have pushed the door wide open for gene editing as a way to pursue not only therapeutic ends but human enhancement on a different scale.

Something is happening as the time of algorithms gains ground and infiltrates everyone's life. It plays havoc with existing timelines, mixing them up and cutting through them. Sometimes it seems as though Einstein's physical time, the space-time continuum that does not distinguish between past and future, has clandestinely entered the social world. It leaves us puzzled, if not confused. Predictive analytics based on algorithms uses *our* past to transform it into *our* future. Living in the *now,* the special present moment with the capability to reconstruct the past through memory and to imagine the future, has gained in prominence. It is now a widely accepted perspective. It serves as the social anchoring point to ease commuting through different time zones. Time travel is no longer confined to the literary genre, with exotic trips into a future and back. It has found an equivalent in everyday life, as we commute between different electronic time zones, set up and controlled by a multitude of algorithmic time regimes set alongside the temporal regimes governed by institutions. Those who are caught in between are like commuters stranded between different time zones.

Some of the spectacular advances of AIs have been highlighted in the hard-fought games played against some of the best human players. It began with Deep Blue, which famously won against chess grandmaster Kasparov in 1997. In 2016 Google's AI firm DeepMind defeated the world's best players in Go in a string of online games. The most recent victory of an AI was achieved in poker, another strategic game, when a team of the world's best professional players suffered a humiliating defeat. It turned out that the AI's main advantage was not the ability to analyze its opponents' strategies. Rather, it was the ultra-rapid learning the AI went through in analyzing the mistakes it had made and correcting them. This shows that an AI can learn faster at a deeper level and is able to apply what it has learned. The question arises what this means for humans and how fast they will learn to maintain control. It takes time to grow and to mature, just as it takes time to learn. Will we be able to keep up and stay ahead of what we have created?

The messiness unleashed by broken timelines can only be overcome by weaving a new tapestry that succeeds in connecting the fast with the slow, *eigenzeit* with life time, the short-term with the long-term, and organic rhythms with the artificially generated. It will also require flexibility and dexterity for institutions to learn how to commute between different time zones and how to read their respective timetables, if they exist.

The hope is to overcome the "cursèd spite" and to get time back into joint—into a form and following temporal patterns that are still to be invented.

Fragmented spaces

Spaces are physical as much as social. They facilitate the formation of social groups, enabling individuals to bond and unite. Or they may hinder them from doing so, especially when access is restricted. Spaces are shaped by the dynamic movements of their content. They are defined by boundaries that may be physical or political, linguistic or social, that always include as much as they exclude. The fragmentation of space is a regularly occurring process when social groups divide. Following a breakup, they may drift apart, settle peacefully in separate spaces, or turn against each other, fighting for a larger share of the territory. Fragmentation raises many questions: when it occurs, how, who is involved, and why.

Following these processes has been on the research agenda of geographers, architects, designers, psychologists, anthropologists, economists, sociologists, political scientists, and historians—practically all social scientists. Just as there are cognitive limits to the number of people with whom one can maintain stable social relationships—"Dunbar's number" sets this limit at 150—size also matters in processes of fragmentation. It is by far not the only determinant, as similarities to natural ecosystems inhabited by plants and various animal species suggest. For humans, additional factors enter. They range from kinship ties to political alliances, from values to economic interests that either bind communities together or drive them apart.

Fragmentation of space creates new kinds of messiness. Fragmentation induces rearrangements before new spatial settlement sets in. In the following I will discuss three areas where such rearrangements are currently under way. These are physical spaces of global mobility, virtual spaces in which cyber-fragmentation occurs, and the risk of a rupture in the imaginary space where science and society meet.

Mobility, as the movement of people, things, and ideas across spaces, has been part of human history from the beginning. Given the past trajectories of our ancestors around the globe, we are all migrants or descendants of migrants. Lately, global mobility has received an unprecedented boost; international air transport has increased by almost 6 percent in 2016 alone, to 3.6 billion passengers worldwide. Airplanes need not only airspace and assigned flight routes but also space on the ground—airports where passengers spend non-negligible hours before departure. In an intriguing ethnographic study, Max Hirsh has used airport planning and design in Asia as an incisive lens for interpreting broader changes in contemporary Asian cities.[5] Cross-border mobility flows of a bewildering mixture of temporary populations—tourists, migrants, and a variety of commuters—need to be managed and coordinated. Shifts in the spatial-temporal continuum of new hypermobile and transient populations also demand that urban planners must accommodate these iterant urban dwellers, many of whom require short- term or part-time residence, in their future scenarios of Asian mega-cities.

Hirsh uncovers the stark contrast between the needs of Asia's extremely heterogeneous flying publics and the fantastical design concepts of airport planners and architects. The glass and concrete architecture, equipped with the latest hi-tech gadgetry, is a modernist dream come true. A spacious environment inside the buildings, filled with global boutiques for lucrative tourism to take off and for expected, but yet to materialize, business profits on prior investment, creates a space ready to welcome a future that has arrived in this part of the world. Yet the experiences of the actual users could not be further removed from this.

The mega-airports in Southeast Asia not only serve tourists and traveling businesspeople but also accommodate millions of migrant workers on their nomadic moves. Airport space, while serving all passengers, is a highly fragmented space. Hirsh unravels the economy that underpins it. He shows the widespread user-driven adaptations, such as machines that sell airline tickets and vendors who, for a small fee, instruct people how to use them. Millions of migrant workers move horizontally between one place that is poor and another nearby that promises improvement. But

Hirsh argues that the disconnect between utopian design and an overtaxed quotidian urban landscape avoids engaging with the bifurcation of Asia's middle class into two distinct groups: one that is global, mobile, and expanding; and another that is local, immobile, and resentful.

In contrast to the "shrinking" middle class observed in the United States and Europe, the growing middle class in many Asian countries poses new societal quandaries. These middle classes articulate new political, spatial, and economic demands, undermining established modes of governance. Between the socioeconomic extremes of the exploitation of migrant workers and the gluttonous real estate practices of transnational elites lie the challenges connected to this "middling" class with its aspiration for increasing participation, hence meddling, also in decision-making. It consists of a moderately educated mass middle class that is neither mired in poverty nor entrenched in elite power networks. Seen through the lens of the fragmented space of Asian mega-airports, a growing conflict between the rooted and the mobile segments of the population can be detected.

People travel not only in terrestrial spaces but also in cyberspace. Every year some 3.6 billion people—nearly half of the world's population—have access to the internet. Coincidentally, almost the same number cross the skies in airplanes. Growing affordability and 70 percent world coverage by broadband will further increase the number of internet users. Cyberspace has become so popular that we spend a good part of our lives in it. But it also leads to the formation of bubbles of like-minded, if volatile, social media communities and platforms where people get together and share information. They are susceptible to confirmation bias. But there is more to it. As algorithms are programmed to disseminate more widely the “likes” generated by pressing a button, the most frequent messages are prioritized to jump the queue. This is one way of giving a boost to hate-based messages.

According to Thomas L. Friedman, the world hit a tipping point in late 2016, when “a critical mass of our lives and work had shifted away from the terrestrial world” to cyberspace. “A critical mass of our interactions had moved to a realm where we’re all connected but no one’s in charge,” Friedman says, raising the specter of anarchy and complete lawlessness.[6] In fairness to the early dreamers who built the internet, their idealism aimed exactly at the creation of such an egalitarian and emancipatory space. Their vision was to connect as many people as possible with as much information and online expertise as possible. They did not envisage that given a certain size, networks would split. They underestimated the possibilities for misuse and abuse that every new technology harbors. We are now discovering what hacking, unconstrained hate speech, the deliberate fabrication of fake news, and targeting specific groups are able to achieve even in liberal democracies. We are experiencing how divisive cyber-fragmentation has become.

It is inevitable that, as they grow in size, social media communities or other aggregate units will break up. The evolution of aggregate ecosystems, whether social networks or polymers, follows a rather precise mathematical form similar to fragmentation-coalescence processes in physics. Identification of such a mechanism suggests, for instance, that on the practical side, small aggregates of terrorist networks can be broken up before they develop into larger, potentially powerful ones.[7]

Another space in risk of fragmentation is neither physical like airports nor virtual like cyberspace. It is an imaginary space in which science encounters society in multiple real and imagined ways. In ancient Greece, the marketplace (*agorà*) was the space for citizens to go about their business and to engage in politics, but it also had places of worship for the gods who protected the city. Above all, the *agorà* was a public space for discussing everything that mattered to citizens, even though citizen status was limited to adult males and excluded women and slaves, who constituted a majority of the population. Despite the admiration later enjoyed by the ideal of Greek democracy as direct representation, it could neither be transferred nor upscaled. We have lost the *agorà*, but something of the idea lives on. In an admittedly idealistic and emancipatory vein, it has become the metaphorical space for free and enlightened discussions where "science" meets "society."[8]

In his opening address at the 2016 Nobel Prize Award Ceremony, Carl-Henrik Heldin, chair of the Nobel Foundation, drew parallels with the late 19th-century world in which Alfred Nobel lived and worked. Nobel's era was one of rapid industrialization and economic expansion. Progressive political ideas about peaceful international cooperation flourished, but nationalism, xenophobia, geopolitical tensions, and terrorism were also on the rise. Hateful propaganda circulated, and anarchists assassinated a Russian czar, an Austrian empress, and American and French presidents. The outbreak of World War I dealt a near-fatal blow to European civilization.

The similarities to today's world are obvious. Scientists continue to surprise us with amazing discoveries, and billions of people around the world have been lifted out of poverty. But dark clouds have formed on the horizon. Terrorists have struck Europe with a vengeance. Millions of refugees fleeing wars and hunger are taxing European institutions and straining social cohesion. Populist movements have emerged, calling for closed borders and new walls. But there may be one significant difference that should concern us. In Heldin's words: "The grim truth is that we can no longer take it for granted that people believe in science, facts, and knowledge."[9]

If half of the world's population uses air transport every year (including frequent flyers) and half are internet users (with overlaps between the two groups), it should give us pause to think that soon all 7.4 billion people on earth will be implicated in the imaginary space of science and society. Whether it is food, water, epidemics and health care, climate change, war, the future of work, or any of the other global challenges that confront humanity today, the scientific-technological order and the social order are mutually interdependent, co-evolving and co-producing the world we live in. What, indeed, will happen if "people no longer believe in science, facts, and knowledge"? If the multiple processes of co-production and co-evolution suffer serious breakdowns? When climate change is denied, officials warn that vaccination is harmful, and facts are considered to be merely opinions? What if science and society start to drift apart?

The possibility of such a rupture is real. Experts have been publicly denigrated, high court judges lambasted as "enemies of the people," and a judge in Washington has publicly been demoted by his president to a "so-called judge." Science has not yet come under direct assault, but the occasion will soon arise. It will be an illusion to think that the firewall between the work of scientists and the political abuse of such work can be kept intact as before. Scientists should ask themselves where the place of people is in the science they do. But the repeated insistence on the legitimacy and sole authority of the scientific community to validate scientific facts as the ultimate approximation of truth is unlikely to carry much weight in a political arena where leaders and their followers have contemptuously turned their back on science.

Facts, including scientific facts, need to be interpreted, and the often arduous process of how scientific facts are identified as such must be made transparent and understood.[10] Nor will "a belief in science" suddenly reappear. Competing belief systems abound. Knowledge may have intrinsic value, but most of the public's expectations are about its practical use. Their interest is in the benefits that come with it. Scientists have to acknowledge that they are part of the "elites" who are accused of having become "disconnected" from "the people." The rejection of scientific knowledge implies a loss of trust, but it is less a loss of trust in science than distrust resulting from the perceived remoteness of the elites from the real day-to-day concerns of people. It has become a political statement as part of a populist agenda. As such, it will have to be met on political ground.

In knowledge-based societies, a paradox arises. Expertise embodied in a broad range of professionals and in vast areas with technoscientific dimensions is in great demand.

At the same time, it has become highly contested, as the lines between expert and lay knowledge are no longer clear. Easy access to information on the internet has widened what people can know, even if their ability to assess the information's reliability and validity has not kept pace. Moreover, people have now become *users* and form part of a huge economic enterprise that depends upon their voluntary and involuntary contributions.

The economic integration of lay experts as users lies behind the political demands for the "democratization of expertise." It seeks to include all the stakeholders affected by decisions with a scientific and technological component. The search for new modes of governance has initiated a flurry of participatory experiments. They range from the open-science and open-innovation movement to various forms of citizen science and to RRI (Responsible Research and Innovation).

Whether these ongoing experiments and initiatives will halt the drift between science and society, or even reverse it, remains to be seen. Citizen science projects involving lay people in research in various ways have produced a particularly dense promissory discourse: a greater democratization of science, better scientific literacy, and the enabling of new scientific breakthroughs.[11] Open access has scored some important advances, but many issues remain unresolved. Open innovation resonates strongly with governments eager to promote innovation. But contributions remain strongly asymmetrical, and incentives may prove too weak or not sustainable. RRI places too much responsibility on individual scientists and too little on institutions. The great heterogeneity of these attempts to involve citizens in various participatory capacities is as much a window into the transformation of science as it is into the transformation of contemporary societies.[12]

Fragmented spaces, whether physical, virtual, or imaginary, raise the specter of sharp polarities, of a possible or already enacted breakdown of solidarity and communication. At the same time, they reflect broader changes and developments that unfold elsewhere and of which they are merely a part. Looking through and beyond the occurring fragmentation offers a view of a particular kind of messiness in which some kind of new coalescence is under way. It mixes what is separated and insists on carving out a new space for what is yet to come. Sometimes the mixture appears to be toxic, but it can also come with the promise of healing the existing fractures.

Ways of world ordering

One of the most hilarious classifications ever was invented by Jorge Luis Borges. This fictitious taxonomy of animals is divided into fourteen abstruse categories, such as "Those that belong to the Emperor; Embalmed ones; Suckling pigs; Mermaids, Fabulous ones; etc.," which are allegedly to be found in the *Celestial Emporium of Benevolent Knowledge*. In the preface to *The Order of Things*, Michel Foucault refers to it thus: "it arose out of the laughter that shattered... all the familiar landmarks of thought—*our* thought... breaking up all ordered surfaces... with which we are accustomed to tame the wild profusion of existing things...."[13] In one stroke it demonstrates the absurdity of categorizing things that exist in one (fictitious) culture and transforming them into the (fictitious) categories of another. If you are looking for a good laugh, it is a superb source.

Yet no culture can do without classification, the taming of "the wild profusion of existing things" that need to be ordered and fitted into a culture's ways of thinking. Anthropologists have delved with delight into the enormous variety of ontologies invented by indigenous communities around the world. In exploring the strangeness of the Other, they remain conscious of their own strangeness, that of Western thinking. Classifications and taxonomies, binaries, and other distinctions; dualisms, monisms, and pluralisms—they all rely on some basic distinctions between similarities and differences that are ascribed to an exterior world and an interior one. Yet the boundaries separating what is seen as part of an external environment and what belongs to an inner realm, be it human or spirits, organic or physical, are as varied as the natural surroundings in which they were generated.

An ambitious attempt to introduce a meta-order into world ordering was made by Philippe Descola who proposed to fit world views into four encompassing ontologies. The different world views and cosmologies assembled from five continents permeate the respective social order that has been constructed in a specific environment and is enacted through these ontologies. Each of the four unfolds along a continuum of similarities and differences along an inner dimension, *interiority,* and an outer one, *physicality,* that are projected into human and non-human entities. In the ontology called animism, for instance, the inner dimension of humans and non-humans is believed to be the same, while they differ in their physical appearance. In totemism, both inner and outer dimensions are believed to be similar, while analogism is marked by both inner and outer differences. Descola's ontologies are supported by an overwhelming wealth of detailed descriptions. They are not mutually exclusive; nor are the communities of practice slavishly bound to them.[14]

The fourth ontology, naturalism, roughly corresponds to the prevalent mode of identification of the modern world. It assumes the coexistence of a single unifying physical nature, in contrast to a multiplicity of cultures. Nature is believed to exist independently of us, based on exterior principles. It implies a counterpart—an artificial and human-made world, capable of transforming the natural world. The dividing line between human and non-human entities is central to modern ways of world ordering, while locating humans within a material, Darwinian continuum. At the same time, in naturalism the human mind remains associated with moral consciousness, the soul, subjectivity, language, the idea of the self, and other characteristics that distinguish all humans from non-humans. Modern science and technology have, however, begun to cross these lines, leading, in Descola's words, towards "a new exciting period of intellectual and political turmoil."[15]

The lens of naturalism allows for better understanding of the scientific, intellectual, and technological processes that enabled a modern world order to be built. While the naturalistic narratives inscribed in their ontological basis still dominate, we begin to appreciate that the amazing discoveries and advances made by the techno-sciences continue to challenge the dividing line between what is considered human and non-human. The rapid adoption of CRISPR, a recently invented gene-editing tool, and the accelerated advances in robotics and AI represent, among other things, a dramatic opening that not only undermines cherished dichotomies but is on the verge of producing newly emerging life forms and ways of living.

The modern naturalistic ontology and the making of modernity are inextricably entangled with the techno-sciences. The world view espoused by modern people underpins their visions of human betterment and of building institutions that can shape society. They have become the driving forces of economic growth and wealth creation. The transformation of the natural order is made to fit the goals and values of the social order, which, in turn, must adapt to fit the order thus created. These are the processes of co-production and co-evolution that link epistemological and political as well as practical questions.

The world ordering of modernity proceeded along two paths. One was conceptual, the other instrumental and technological. Conceptually, modern world ordering heavily relied on clearly delineated dichotomies. They tend to be mutually exclusive, with not much left in between. Modern world ordering relied on mathematics, and with the rise of the nation-state, quantification became an efficient means of governing. Numbers, counting, and accounting based on quantifiable units of measurement were also used in "Making Up People" (in Ian Hacking's phrase) through the categorization they helped to construct. Statistics imposed order on a wide range of economic and social arrangements. Modern institutions, from schools to hospitals, from the state bureaucracy to the military, were built to embody a strong nation-state, taking pride in its efficiency and ruthless exercise of power. Hobbes's metaphor of the state as a machine constructed to fulfill the purposes of governing endured in the view of a modern democratic state as a self-regulating machine with its checks and balances.

The neat definitions, clear-cut categories, and well-defined dichotomies of modern world ordering leave little room for "boundary work." This is a term describing the negotiations and possibilities for consensus to arise from otherwise different positions and perspectives. It feeds on ambiguities that enable different interpretations and approaches.[16] Instead, modernity insists on boundaries that sharply separate reason from emotions, subjectivity from objectivity, and facts from fiction. These were hard-won distinctions. They were the outcome of heated debates and of power struggles that pitted traditional forces against those seeking to impose a modern social order.

Nowhere was this more in evidence than in the political realm. Political thought provided legitimacy and authority for the collective political imagination on how to govern. Political life is unthinkable without these imaginaries. As Yaron Ezrahi puts it: "A democracy, like any other political regime, must be imagined and performed reasonably well in order to exist." The political imaginaries that acquire the power to regulate and shape political behavior and institutions are necessary fictions. Ezrahi shows that it is precisely this dichotomy between facts and fictions that served the projects of modernity, while it also obscured the unique potential of the imaginary to be both fact and fiction. In many respects, the imaginary is the hidden shaper of politics.[17]

Like other powerful dichotomies that gained enormous significance for the modern way of world ordering, the distinction between facts and fiction became self-confirming. It reinforced the conceptual grid underpinning the modern world and reaffirmed it through efficient performance management. The political imaginaries evoked by modernity expanded their hegemony well beyond Europe. They conquered the cultural imaginaries and succeeded in separating modern people from those perceived to be backward and living in pre-modern conditions. The lure to become modern proved irresistible. European colonial expansion, which carried horrific costs and consequences for those who were supposed to be lifted out of their inferior state, was backed by the particularly hideous dichotomy between civilized and non-civilized people and nations.

Modern world ordering came with a timeline that none of other ontologies knew. It was the idea of an open future. Progress was perceived as a straight path of continuous economic growth, with the techno-sciences pursuing the future by promising further advances. Artists, given their prescient sensitivity, saw it somewhat differently. Baudelaire famously defined modernity as the fleeting, the transient, and the contingent. However, these features determined only one half of his vision of modern art. The other half remained anchored in the ideal of eternal beauty. It was immutable. Baudelaire's idea of modernity reverberates to this day, when artists celebrate the ephemeral and transient. Artists were also at the forefront of blurring the previously strict boundaries between high and low culture, infusing art with the vital energy that springs from creative mixtures.

The arts were not the only domain where the meticulously drawn boundaries that fit the modern grid and trajectories laid out for the future of modernity began to give way. The late 19th century gave birth not only to nationalism but also to liberal and emancipatory ideas and movements, to imagined communities bound together not only by national identities but also by adherence to a rational dialogue between citizens. This led to thinking that wars could be overcome through rational decision-making.[18] The big shock came with World War I, when Western civilization suffered a stroke from which it never completely recovered. The idea of progress, linking technological prowess with social and even moral progress, received a fatal blow.

So what went wrong? What happened that was able to undermine well-defined dichotomies and clear-cut grids that were not only performed in thought and action but also expressed in the material, built-up modern world full of impressive achievements? Why did many of the control mechanisms built into modern world ordering fail to prevent their erosion? Many answers have been offered to these and related questions. One obvious answer is that appearances can be deceptive. Behind the orderly surfaces, unruly and contesting counter-forces are always at work. The stupendous achievements of modernity cannot eclipse the fact that under the banner of modernity, horrendous crimes were committed, often in the name of ideologies and publicly claimed virtues. Order is always under threat, and disorder lurks in the background. The book *Seeing Like a State* summarizes where the centralization of power and excessive control, which have been characteristic of attempts to maintain order at any price, can lead.[19]

In this sense, as Bruno Latour puts it, "we have never been modern." There has always been duplicity and hypocrisy in the West and elsewhere. In the arts, such tensions can give rise to bursts of creativity. When it comes to institutions and social structures, the intentional jostling between what is real and what is make-believe has much more serious consequences. Once the main boundaries between the spheres of influence of the state and the market started to give way, the global winds of neoliberalism led to the loosening or abandonment of many regulations and state controls. Transnational corporations and global markets took over, putting considerable strains on national governments with limited sovereignty and on the quality of the functioning of state institutions.

While there are other powerful converging tendencies at work, the biggest thrust towards undermining modern world ordering emanates from within. It is one of the great achievements of the naturalistic ontology providing the framing for modernity that pushes towards modernity's implosion—the slow but inescapable takeover of the natural order through the artificial and human-made. As new technologies thoroughly reshape the world we live in, what makes us human is up for continuous redefinition. Together with market forces acting globally, we are undergoing another major transition. We live in a time of in-between, experiencing the "pregnant widow syndrome." The old never vanishes completely. It lives on and is transformed, while the new is not yet fully visible and growing. It is immature but already kicks, wanting to release the energy inside it. Technologies are never neutral or innocent. They are designed with specific purposes in mind but may be deployed for completely different uses. They come loaded with promises, risks, and even threats. We have indeed entered a phase of intellectual and political turmoil, as Descola dryly observed. The human and the artificial created by humans have to find ways to accommodate each other, with in-betweenness still to be shaped.

The child in the making, visualized in the countless images that computerized technology continues to spawn, is the child of the enormous transformative power of computation. It rests in algorithms, in the rules and programs that enable connectivities to be established at high speed, covering domains hitherto impervious to such connections. Its potential is unleashed in the labor market, impacting the future of work while leaving open the question of who will pay for those who are unlikely to get into, or back into, the labour market. Computation has successfully found its way into the life sciences, where huge amounts of data are available to permit insights into patterns of regulatory pathways that hover above health and disease. Correlations come to the fore between genetics and lifestyle, highlighting epigenetic phenomena. The mechanistic worldview of science, successful as it has been under modernity, is yielding to a more organic world view. Like evolution, it insists on trying out new combinations of diverse components. It is definitely not about purity but more about bastardy.

The newly emerging computational world ordering overthrows the foundational dimensions of interiority and physicality on which Philippe Descola's naturalistic ontology rests. Many of the previously fixed lines of demarcation between the natural and artificial, the human and the nonhuman, are challenged as speed and efficacy accelerate. New combinations arise that are far more complex than the simple hybrids that always had their place as anomalies under the old world ordering. Recently invented genomic tools like CRISPR have brought gene editing closer to becoming a routine operation for therapeutic intervention. Part of the insatiable quest for human enhancement, they are transforming the body into an experimental testing ground for new combinatorics between organic and non-organic component parts. We are in the midst of an overwhelming world reordering, with AI taking over many domains that previously were thought to be impervious to machines replacing humans. In their stead new interfaces arise, reordering and repurposing humans in the human-made world they continue to make and change.

We are not there as yet, in this brave new world that will turn out different from both our worst fears and our most ambitious dreams. In the tension between old and new and its broad reach, we should be wary of the boundless techno-optimism that celebrates the singularity that is supposedly approaching rapidly, as well as of the apocalyptic scenarios that warn of a rapid loss of control in the face of AI. The new will become absorbed, and the old will prove more resilient than we now imagine. The most effective form of world ordering so far—modernity—has not yet been defeated.

Reordering messiness

The ordering of the world is an ongoing process. It follows, contradicts, and responds to an unruly reality. It works through concepts as well as material artifacts that relate ideas to the surroundings in which people live. For the first time in human history, the powerful scientific and technological tools at our disposal enable us to see further into the past and, to a limited extent, into the future. They render visible what used to be invisible. But there is a price to pay that concerns us directly. Humans are no longer mere observers who watch from the outside. We can no longer claim to be innocent and neutral bystanders. To paraphrase Thomas Nagel, there is "no view from nowhere." We and our actions are now an integral part of a world by adapting and reacting to its complexity. The hard fight for the attainment of scientific objectivity separating facts from values and intersubjectively validated findings now has to accommodate subjective intuition and emotion. This contributes to messiness.

At the start of the 20th century, science was marked by a clear subject/object distinction. Now new spaces between objectivity and subjectivity open up. Where sharp polarities once reigned, we now tend to see a continuum. This holds for diseases like autism or schizophrenia as much as for social phenomena like living in times of war or peace. Likewise, the once-promised certainty of science had to make room for uncertainty, at least in the sense of a better understanding under which conditions knowledge said to be certain holds. The unlimited power to predict turned out to be limited in ways that indicate the constraints for prediction, moving from probabilities to the probability of probabilities. Computers and computational thinking have enabled us to harness randomness, and contingency has replaced absolute certainty.

Obliterating once clearly demarcated dichotomies, such as rational and irrational, has nowhere been as much in evidence as in economic behavior. The vast range of conceptions about the behavior of "rational man" has deftly been undone, as exemplified by the work of two well-known figures from behavioral economics, Amos Tversky and Daniel Kahneman. Together with other colleagues, they showed human judgment to be inferior to simple models. The rules of the mind work very differently from the rational behavior attributed to and expected from it.[20] But this is only the human equivalent of the tip of an iceberg. Seen through the lens of complexity science, the economy is a complex system that starts to look and behave differently. It can no longer be thought of in terms of striving for a mechanistic equilibrium or of "solutions" in precise mathematical terms. Instead, it becomes a system in motion that perpetually "computes" itself and thus constructs itself anew. This new framework emphasizes contingency, indeterminacy, sense-making, and openness to change. The focus is on a set of emergent phenomena that form a pattern that is changing and inducing further change. Similar to other fields of science, economics is becoming more procedural and algorithmic, more Turing-like and less Newtonian.[21]

Human actors are an integral part of most complex systems that are presently explored.[22] Their emotions, foibles, and cognitive biases can no longer be kept outside. In the abstract language of complexity science, they are the multiple elements that adapt and react to the very patterns these same elements create. The diverse networks that constitute contemporary societies—from finance to terrorism—emerge from social interactions. For instance, the study of natural ecosystems and of biodiversity can no longer be separated from human interventions and their impact on the natural system of the earth. The hypothesis of the Great Acceleration posits a strong correlation between them that appears for the first time around the 1950s. Changes in human production and consumption, like energy use, telecommunications behavior, and GDP, are reflected in changes of greenhouse gas emissions, ocean acidification, terrestrial biosphere degradation, and rates of extinction.[23] Together, they constitute the complex system of life on this earth and the vulnerability, as well as the resilience, it displays.

Of the many blurring boundaries in flux, the most dynamic is the one separating the natural from the artificial. Its potentially explosive power is grounded in the naturalistic ontology framing the scientific worldview. It pushes for the continuing redefinition of what makes us human. Technologies designed to form new parts of bodies and minds recombine and repurpose what it means to be human. This requires that social scientists and the humanities be brought into a field dominated by engineers, mathematicians, physicists, and computer experts. How do we account for the fact that expectations create the outcome that the forecasts are trying to predict? And while it may suffice to look at aggregate patterns formed by individual behavior and preferences, can we shut out the individual completely? And where does society, which is more than the aggregate of individuals, enter the picture?

Bringing people, represented by artificially created agents, into simulation models and thus into a simulated world in order has unintended consequences. Whether the models are about stock markets or urban transport systems, commercial fishing or the formation of terrorist networks, the artificially created world is not isolated from the real, social world. It interacts with it, precisely through the predictions it makes. This often happens in unforeseen ways. As is well known, risk modeling in the financial world carries its own risks. Herd behavior may follow forecasts. Self-fulfilling prophecies are a frequent social phenomenon. Many professions are already replacing with automated procedures what humans did before. These procedures are able to combine different kinds of data and data from different domains, with different visualization practices, programmed for different questions and purposes. But the social world, with its governance structures and bureaucracies, with corporate interests and transnational flows of talent and finance, is still there. It continues to exert a strong pull while being pushed at the same time. Whatever hybrid forms of social organization will emerge, whatever blending between old and new ways of world ordering will be the outcome—agent-based modeling will need linkages with the social world whose problems it promises to resolve.

Such interlinkages raise some interesting questions. Could the very interactions between different elements of a complex system consisting of human and non-human elements ever become the primary units of agency? Could these new units of agency replace notions of an autonomous self and of a self-contained identity? If so, the social order would no longer consist of hierarchies with persons in differential positions of power and submission, but of interacting and coupling agent-units that easily switch between bottom-up and top-down modes depending on the task they fulfill. This would not mean that individuals and their actions would be pre-programmed and set on automatic pilot. Nor would it mean that the old dream of an egalitarian society would finally come true. It would only raise the intriguing possibility that new forms of social life and togetherness could emerge, navigating an in-between space of cherished concepts like social justice and equality and concepts like symbiosis and parasitism.

These are speculative thoughts that may never materialize. But they should prepare us for the unexpected, in the guise not only of technological innovation but also of the new combinatorics between human agency and technology that is likely to emerge. In this context, design would assume an even more prominent role. It is concerned not only with how things are but with how they ought to be. Design devises artifacts that are made to attain goals. The engineering approach that underlies programming and design and which drives the ongoing reordering in the techno-sciences in a digitalized and combinatorial space has made impressive strides in meeting the organic evolution of life. Progress in the life sciences has benefited enormously from adopting and integrating a computational and quantitative-based approach. Something similar is yet to come for the social sciences.

The world ordering based on computers and computational thinking is not self-contained. It spills over into the social world. But there is also a spillover from nature that surrounds us and still acts as a powerful force. Human intervention in the natural environment has not yet tamed volcanos and earthquakes. Tsunamis can still wreak immense damage, especially when near nuclear power plants. Recklessly induced anthropogenic climate change leads to melting in the Arctic and ocean acidification. And while many diseases have successfully been brought under control, pandemics and outbreaks of new diseases are likely to continue in the future, simply because we live in an interconnected world. As connections intensify, so does the vulnerability to "black swans": rare and unexpected events with a huge impact. The ability to predict grapples with many cognitive biases. Resilience of and within societies is in high demand, as the ability not only to return to a normal state after disruption but to do so with a strengthened societal immune system.

In the 1930s the German writer and philosopher Walter Benjamin analyzed the effect of the invention of photography and filmmaking on works of art.[24] The camera enabled artworks to become dislocated. They were freed from being fixed in one location and could now, thanks to their mechanical reproduction, be inserted into different contexts. They lost their originality but gained ubiquitousness. In this tradeoff, their meaning changes according to the context in which they are displayed, used, or misused. Thus Mona Lisa appears on hundreds of ads and gadgets, joined by many other artworks now appropriated in a Disneyfication of culture. There is no "out of context" anymore when context can be made fluid at will. When context becomes interchangeable, viewers have to invest it with their own meaning.

On a larger scale, something similar happens for the newly emerging patterns formed through interactions and recombinations of unlikely elements. They take on different meanings when combined with other elements or units of analysis. In the language of Stuart Kauffmann, innovative ideas cross the border from the actual to the "adjacent possible." They emerge from the interplay between the actual and the possible, but each new idea changes the landscape of future possibilities in the adjacent possible.[25] But what exactly is the adjacent possible? How can we explore its possibilities and the potential of the messy new combinations it might offer? Messiness refuses to be subsumed under a pessimistic, optimistic, or possibilistic stance. It invites us to work our way through the bewildering dynamics of a new world ordering.

Here is where historical precedents enter. Seductively, they offer themselves as guides for a future that is unknown. In Europe, two of the most powerful intellectual and social movements were the Renaissance and the Enlightenment, with the origins of modern science strongly interwoven in between. They no longer appear today simply as embodied in the genial minds of a few well-known figures. Rather, they represent networks with a few hubs in a vastly extended and diverse European landscape, often linking obscure figures in hardly known and remote places. Despite political and religious obstacles, ideas, instruments, and skills that invigorated them easily crossed borders. This networked interconnectedness constituted the fertile ground of the adjacent possible. It is no wonder that dreams of a possible revival have led to notions of a New Renaissance or an Enlightenment 2.0.

But history does not repeat itself; nor can intellectual movements be engineered. Fixing our gaze on the adjacent possible that surrounds today's computational technologies and the procedural thinking that comes with it should alert us to further explore what possibly lies adjacent. But when it comes to the legacy of modernity, a stark warning needs to be heeded. We should be extremely wary not to replace the centralized, rigid dichotomies on which the omnipotent fantasies and the horrendous consequences of *Seeing Like A State* were based with another omnipotent fantasy: *Seeing Like An AI.* Its impact will be even more devastating and far-reaching.

Exploring the adjacent possible implies aligning with the forces of hope. This means accepting contingency and the openness it entails. Messiness can release the creative energy that comes with the diversity of experimental mixing and combining. Computer modeling is blind if it does not include a vision of society and how we can achieve better ways of living together. The social sciences and humanities have to find their own adjacent possible in the computational world ordering in the making. Together, we may learn to do what complex adaptive systems are all about—adapt and react to what we create—while striving to bring about a better society.

Epilogue

In the dark year of 1941, when it felt as if the civilization of the last few centuries was fast disintegrating, the well-known biologist who discovered epigenetics, C. H. Waddington, wrote a short book on the role of science and scientific values in resisting a Nazi attack. He feared that Britain was likely either to be invaded or overrun like France, or to make some kind of sellout arrangement. The book was a passionate plea for science to include the "participation of all aspects of the human personality, not only the rational, logical mind, but emotions and unconscious thought-processes as well." Science, he asserted, has something to say about values, attitudes toward fascism and communism, religion, racism, world peace, the arts, and scientific culture. In somewhat old-fashioned language, he concludes: "Prim science has so far neglected to confess to the world that he has begotten such an offspring... on the harlot Humanities; but the infant culture is beginning to peep already—in its bastardly vigour lies the only hope for an heir worthy of the civilisation of the past."[26]

Messiness, it seems to me, is full of this bastardly vigor.

Endnotes

1 Alison Gopnik, *The Gardener and the Carpenter* (New York: Farrar, Straus and Giroux, 2016); Raghavendra Gadakar, "The Universe: Which Tools to Understand It" (2016), DOI:10.16943/ptinsa/2016/48860.

2 Tim Harford, *Messy: How to be Creative and Resilient in a Tidy-Minded World* (London: Little, Brown, 2016).

3 Ulrike Felt, "Under the Shadow of Time: Where Indicators and Academic Values Meet" (2017), DOI:10.17351/ests2017.109.

4 Marilyn Strathern, "Comment on Fochler and de Rijcke's 'Implicated in the Indicator Game? An Experimental Debate'" (2017), DOI: https://doi.org/10.17351/ests2017.157.

5 Max Hirsh, *Airport Urbanism: Infrastructure and Mobility in Asia* (Minneapolis: University of Minnesota Press, 2016).

6 Thomas Friedman, "Online and Scared," *New York Times*, January 11, 2017, https://www.nytimes.com/2017/01/11/opinion/online-and-scared.html.

7 Neil Johnson, "New Terrorism Reveals New Physics," *APS News*, November 8, 2016, 8.

8 Helga Nowotny, Peter B. Scott, and Michael T. Gibbons, *Re-Thinking Science: Knowledge and the Public in an Age of Uncertainty* (Cambridge: Polity, 2001).

9 Carl-Henrik Heldin, "Opening Address, Nobel Prize Ceremony" (2016), https://www.nobelprize.org/ceremonies/archive/speeches/opening-2016.html.

10 Ludwik Fleck, *The Genesis and Development of a Scientific Fact* (Chicago: University of Chicago Press, 1979; first published 1936).

11 Bruno Strasser et al., *"Citizen Science"? Rethinking Science and Public Participation* (University of Geneva, 2017).

12 J. Chilvers and M. Kearnes, eds., *Remaking Participation: Science, Environment and Emergent Publics* (London: Routledge, 2016).

13 Michel Foucault, *The Order of Things: An Archaeology of Human Sciences* (London: Vintage; reissue edition, 1994; first published 1966).

14 Philippe Descola, *Beyond Nature and Culture* (Chicago: University of Chicago Press, 2013).

15 Philippe Descola, "Beyond Nature and Culture: Radcliffe-Brown Lecture in Social Anthropology, 2005," *Proceedings of the British Academy* 139 (2006): 137–155.

16 Susan Leigh Star and James R. Griesemer, "Institutional Ecology, 'Translations,' and Boundary Objects: Amateurs and Professionals in Berkeley's Museum of Vertebrate Zoology, 1907–1939," *Social Studies of Science* 19 (1989): 387–420.

17 Yaron Ezrahi, *Imagined Democracies: Necessary Political Fictions* (Cambridge: Cambridge University Press, 2012).

18 Wolfram Kaiser and Johan Schot, *Writing the Rules for Europe: Experts, Cartels and International Organizations* (Houndsmills: Palgrave Macmillan, 2014).

19 James C. Scott, *Seeing Like A State: How Certain Schemes to Improve the Human Condition Failed* (New Haven, CT: Yale University Press, 1998).

20 Michael Lewis, *The Undoing Project: A Friendship that Changed the World* (Allen Lane/WW Norton, 2017).

21 Brian A. Arthur, *Complexity and the Economy* (Oxford University Press, 2015).

22 Stefan Thurner, ed., *Exploring Complexity*, vol. 3, *43 Visions for Complexity* (Singapore: World Scientific, 2016).

23 Owen Gaffney and Will Steffen, “The Anthropocene Equation,” *Anthropocene Review*, DOI: 10.1177/2053019616688022. First published February 10, 2017.

24 Walter Benjamin, “The Work of Art in the Age of Mechanical Reproduction,” in *Illuminations* (New York: Schocken Books, 1969; first published 1936).

25 Stuart Kauffman, *At Home in the Universe: The Search for the Laws of Self-Organization and Complexity* (Oxford University Press, 1996).

26 C. H. Waddington, *The Scientific Attitude* (London: Hutchinson Educational Ltd., 1968; first published 1941).

Eigenzeit. Revisited

My book *Eigenzeit* was published in the memorable year 1989. It is a very personal book whose growth at the time followed the temporal rhythm of my life. In writing the book, I was aiming for a social-science-based diagnosis of current shifts in the meaning of the concept of time, of exposure to and experience of time. I wanted to report on the contemporary experience people had with time and to analyze the conflicts emerging from it. I observed how the boundaries between private and public time began to blur in everyday life and was fascinated by the gradual absorption of the category of a distant future by what I called the extended present.

Everywhere I looked, I stumbled upon technologies that alter our perception of time in the most immediate and visible way: the new information and communication technologies. *Eigenzeit* set out to demonstrate the qualitative changes in the individual perception of time and the corresponding experience in their manifold connections to the structuration society imposes on time. Technologies played an important role throughout, but the same was true for the globally disruptive, virulent economic and political processes that added a new dimension to the nexus between power and time.

Why did I choose time as a topic? It all started early in 1973. At that time, my then considerably younger self underwent an abrupt transition into a new phase of life. From a hectic professional routine compounded by a turbulent private situation, I found myself transported from one day to the next to serenely idyllic Cambridge for a sabbatical. How come, I asked myself, some people have so much time and others so little? Why are some able to blissfully concentrate on doing one thing at a time, while others have to juggle a multitude of tasks, constantly on the verge of exhaustion? It is fairly obvious that what was on my mind was, above all, my own situation.

Although it was not part of my original plan, I soon settled on seriously studying social time. The renowned anthropologist Edmund Leach was then provost of my host institution in Cambridge, King's College. He was thoroughly familiar with a wide range of different conceptions of time embraced by indigenous people in the most remote parts of the world, who were thus bringing human existence into harmony with the universe, nature, and their specific social order. We talked about the meaning of cyclical time and about the transformation myths of tribes in New Guinea. We even moved into linguistic territory, discussing, for example, the way verbs were being displaced by nouns in Western societies. Leach generously invited me to use his own private library. Thus I immersed myself into reading anything I could lay my hands on that was related to time.

Soon I learned of the existence of the International Society for the Study of Time (ISST). This society was to hold its next conference in Japan in July that year. Then, as now, its members came from a wide range of academic disciplines—physics, linguistics, musicology, sociology, sinology, you name it. The great—and indeed the only—topic they had in common is time in all its fascinating interdisciplinary multitude. I decided to submit a paper and set off for Lake Yamanaka at the foot of Mt. Fuji. In keeping with my budget, I used the Trans-Siberian Railway and later an East German freighter that took passengers on board. On the borrowed typewriter of the captain, I made the final revisions of my paper on the relationship between different structurations of time and time measurement.[1]

Time has stayed with me as a topic ever since. At the conference I quickly learned more about the different disciplinary approaches to time and encountered interesting people. More conferences were to follow as the years went by, and at a later stage I was even elected president of the Society. I remained in touch with the founder of the ISST, J. T. Fraser, until he passed away. In the meantime, my academic career had progressed. My little daughter took pleasure in her mother's exotic interests. She asked me clever questions such as "What is time?" and "Is time the same everywhere?" And she was delighted when she discovered the ambiguity of the English phrase "What time is it?"

Two temporal islands, one month in 1987 and another in 1988, made it possible for me to write *Eigenzeit*. There was no hurry, no deadline to meet—indeed, there was not even a publisher. The book had all the time in the world to grow and to weave together scientific and personal threads of time.

A retrospective distance of three decades creates sufficient space and time to ask new questions. What is left of the late-1980s conception of *eigenzeit*: a longing for the now and the desire to have more time available to oneself? What are the social and scientific-technological developments, with their host of expectations, anxieties, and projections of the future, that have moved to the foreground today? How has the relationship between life time* and *eigenzeit* changed? Does a political dimension concerned with issues of time and temporalities, which seemed so pressing then, still exist today? How do power and time relate to each other at present?

* "Life time" refers to the whole of one's temporal existence—how one lives one's life.

In retrospect, it is, above all, the continuities that stand out, perhaps because they seem to verify the anticipatory analysis. This includes the phenomenon of acceleration, which I had seen as closely linked to the dynamic of scientific-technological development and its economic impact. In this vein, I had written: “While in the phase of industrialisation it was above all the equation of time and money which resulted from the industrial capitalist logic of production and made time a scarce commodity, time is now being speeded up itself: it is becoming accelerated innovation.”[2]

Today, innovation is omnipresent. It is the determined and robust anticipation of the future by the present, supported by the co-production agreement that science and technology have entered into with a society that has eagerly concluded a pact with the new. Innovation is anchored in political rhetoric. It has become a key term denoting a bet politicians are making on an increasingly fragile future, in the belief that they have discovered a failproof mechanism that will enable us to weather any crisis.

The acceleration of innovation has become a matter of ever-greater urgency. The problems created by the loss of jobs owing to the unstoppable advance of technological automation need to be solved—by more innovation. The solution seems to lie in incrementally greater doses of innovation, and it is indeed innovation that is being counted on to create new jobs and to open up new possibilities. The radical changes that computerization and digitalization have wrought over the last few decades in the economic, social, and political fields demand additional radical measures. In practice, however, it looks more like a stumble into a future in which promises of innovation are interwoven in very uneven ways. In the end, innovation must somehow combine the old with the new in unforeseeable ways. Innovation consists above all of recombination.

Having said this, it cannot be denied that the phenomenon of acceleration is one of the most far-reaching consequences of the changes brought about by science and technology. Once again, time seems to be out of joint. What so many people find disturbing today is the speed at which modern life and their own lives are being played out.

Most of us feel alarmed by the dizzying speed at which the world is hurtling forth, a speed that far outstrips our biological, cognitive, neural, and mental capacities. The result is a deep-seated sense of unease and massive stress. This can be countered either by timid defensiveness or by the compensatory opposite, the decision to go with the flow, where the evanescent now becomes the last straw to hold on to in a present that has lost orientation.

What becomes visible here is one of the many lines of conflict resulting from the collision of different temporal regimes. Acceleration, however, is triggered not only by technology's electronic speed. It radically surpasses all human capacities, so much so that attempts at adaptation are doomed to irrelevance. What is lacking is efficient time transformers, the institutional and social arrangements and rule systems that work towards the integration and assimilation of different temporal horizons, speeds, and regimes.

It follows that it would be rash to simply equate the acceleration we experience with technological acceleration. Hartmut Rosa, in *Beschleunigung und Entfremdung*, quite rightly distinguishes between technological, social, and cultural acceleration. In part these processes cause one another, in part they are mutually contradictory. Technological acceleration is visible and palpable for all; it can be measured and is easy to pin down. Social acceleration makes visible the changes that take place in a society. It is often twinned with a loss of stability of institutions, and building new ones takes time. The third characteristic is cultural acceleration, which is reflected in our awareness of the rapid pace at which we are living our lives.[3]

Yet another recent book deals with the paradox of how time pressure continues to build despite the use of time-saving technologies. Its topic is the interlinkage between technological, social, and cultural acceleration. Technologies are supposed to make life easier by streamlining processes and boosting efficiency. However, rather than becoming more plentiful, time is compressed and becomes ever scarcer. In *Pressed for Time*, Judy Wajcman identifies digital capitalism as the root cause of the increase in acceleration. She makes it quite clear that the process of technological innovation always offers a plurality of options. She adduces a wealth of examples to show the different uses people make of digital technologies. The processes by which we construct social reality with the help of digital technologies will not run their course for some time to come.[4]

I fully agree with the diagnosis that the acceleration we experience is not the direct result of the technologies whose working speeds are so radically different from ours. But when I speak about the dynamic of accelerated innovation, it is not only these technologies I have in mind. That dynamic is generated through the multiplicity of social interactions that are built into, facilitated, and mediated by the technologically new production processes and products. Their effect, and the built-in contradictions, are to be found in the networks that are both socially condensed and geographically expanded. They contribute to the interactive game by creating ever-changing social and economic expectations. This gives rise to the temporal complexity that overwhelms us.[5]

The linear sequence of working steps and procedures has been replaced by the omnipresent expectation of the capability to master simultaneity. Prompt execution in real time is taken for granted. As an organizational principle, “just in time” is being superseded by Industry 4.0, which is currently spreading to more and wider areas. Work and everyday life are being shaped by project-like demands. The organizational preconditions for this stem from the potential for recombination, which in turn builds on the fact that many more and more diverse interfaces are available. This is the environment that is indispensable for the temporal and spatial flexibility demanded by markets in the production of goods and services as well as in the workplace. Simultaneity seems to pervade everything.

Seen from this perspective, the experience of acceleration results from the multiplication of expectations that we want to—and presumably have to—meet. It is therefore not the technologically enabled greater speed as such that troubles us. The acceleration we experience is a consequence of the complexity arising from the dense economic and social interconnections as parts of a system that is enabled, boosted, and expanded by technology.

This also applies to the acceleration of natural growth processes that Bernd Scherer has highlighted. Biotechnological regimes are superior to traditional breeding methods to an extent that was unimaginable in the past. In increasingly sophisticated ways, nature is being duped into stepping up her pace. In a convergent development, new financial and business models are created to open up new markets for products maturing ever more quickly. Even the obsolescence traditionally built into industrial production is becoming outdated. It is automated.[6]

What these trends might mean in the future for that only true and rare resource, time, has been outlined by Jacques Attali in a dystopian vision of the 21st century. As it takes less and less time to produce and to market goods, time use shifts from production to consumption. We become the time slaves of consumer goods and services. We have to allocate time to use what we are being offered—be it transport, communication, entertainment, or access to information, which needs to be downloaded and integrated. Does this sound familiar?

Making a selection from this glut becomes more and more time-consuming. Only slowly do people realize that they will never have enough time to learn all they need to know to remain "employable." The markets react to this situation by offering material and virtual goods whose use is ever more time-consuming. Part of the deal is the illusion that the time will come when these commodities will stand us in good stead—as if death could be postponed by activities that are as time-consuming as possible. In the end, everyone realizes that time is the only true and rare resource: no one can produce it, no one can sell it, and no one can store it.

This ushers in the rule of the "hyper markets," a market of markets. States are replaced by markets and disappear completely. In a countermove to the rise in life expectancy, growth processes are shortened. Babies are born earlier; children learn to speak more than one language at an earlier stage and more quickly. All vital functions are speeded up and cut back—sleeping, making love, eating, learning, playing, making decisions. The great crisis, which in Attali's view is already upon us, is about to kick in. And in the far distance beckons the utopia of a "hyper democracy."[7]

The experience of acceleration as a key component of modernity and its metamorphoses are going to be with us for a very long time. What is striking about this foray into the recent literature is that one phenomenon is consistently sidelined that I would highlight and prioritize in "Eigenzeit. Revisited": the accelerating destruction of our natural environment.

From the felling of tropical rain forests to the loss of biodiversity; from the increase in catastrophic droughts to extreme meteorological situations even in our own temperate zone; from increasingly scarce water resources and shortages of foodstuffs to the challenges posed by the rapid spread of megacities—all these phenomena adhere to a logic of acceleration that has set humanity on a seemingly irrevocable collision course with its inherited natural environment. While nature has been subjected to change by *homo faber* from the very beginning, only now has human intervention reached a scale where it jeopardizes the very survival of the human species.

Time is running out. Will we succeed in a race against time to ward off imminent collapse? Open up sufficiently productive sources of renewable energy soon enough? Supply sufficient quantities of water and food for a world population growing faster than expected? And will it be possible for science and technology to speed up the pace of cultural evolution so that life on earth remains worth living—or will this lead to side effects resulting in even more destruction?

Unlike nature's biological evolution, which unfolds in a long-term perspective and has almost infinite timespans for trial and error, cultural evolution progresses at an accelerating pace. Above all, we are the ones that produce the required variation and act upon the selection. We are the ones that cause the cultural equivalent of biological mutations. These are largely not left to chance but are the direct effect of our interventions. But what do we know about the unintended consequences of human action?

It is no longer only the familiar time scale of human history that will decide this race. We are up against time scales of radically different orders of magnitude. This is the first major revision that I would like to introduce to "Eigenzeit. Revisited."

A second revision is at least as significant and momentous. It is the acceleration of the multiplication of available data and, even more importantly, the growth of the capacity to store data, to process them, and to make them available for multiple and different uses. Today's world is shaped by big data.

Digitalization and computerization have created global networks that are being used by states and their secret services, by the financial world, by criminals, by science and industry—by all of us. What makes this possible is algorithms. These are mathematically formulated procedures for the solution of specific problems in the form of highly sophisticated computer programs. They are the *dei in machina*, invisible and therefore all the more efficient. Algorithms are the soft underbelly of complexity.

The speed with which data are processed and recombined far exceeds the capacity of human perception. All we can do is register their effect. What bothers the average user is less the speed than the loss of control, the fact that all data, once they have been gathered, entered, and stored, can be called up at any time and put to new and different uses.

Storage is almost for eternity—or for that portion of eternity that is left to us. Privacy increasingly turns out to be an illusion. And while legal battles are being fought over the "right to be forgotten" on the internet, the feats that the latest information and communications technologies are capable of and what we allow them to do with us have a profound impact on our experience of time. This leads to the next question: how have our perception and our experience of time been altered? What has become of our longing for the now?

Global simultaneity experienced in concrete terms, and its economic and technological impact, which I described in *Eigenzeit*, has expanded even further in the meantime. "Whoever governs simultaneity controls the temporal dependenc[i]es derivable from it. Yet despite the attainment of ever-higher speeds in the networking of information technology, despite the continuing expansion of the technological infrastructure which installs simultaneity, it remains an illusion."[8]

The illusion resides in the assumption that technological simultaneity will do away with global inequalities. Even though it has been possible to lift millions of people around the globe out of the worst poverty, there are still countries where the number of mobile phones exceeds that of toilets. The epidemic caused by the Ebola virus would have run its course with a far less dreadful toll in human lives if Africa's health care system had had a sound infrastructure. And Europe is currently experiencing a confrontation with refugee flows that are no less than a desperate expression of the non-simultaneity of human existence.

This is the great temporal contradiction we live in. Technologically induced simultaneity leads to a homogenization of the perception of time, while social asynchrony is growing in proportion to the gap created by inequality.

Participation in global communication demands both a standardized and an individualized experience of time. This is a precondition for synchronization. Time is split into small, standardized units to be replicated and individually reassembled at will. The result is a short-term perspective that guarantees flexibility. It is key for the exchange of information as the primary mode of communication.

But what information are we talking about? What is its content? In the age of big data, the concept of information has been considerably broadened. Anything that can be processed as data becomes information. Big data pervades countless aspects of the workplace and of everyday life. It eliminates the boundaries between private and public, between what is classified and what is released for public consumption. It creates links between the most diverse areas where human activity and the events resulting from it leave traces.

Data are indifferent. They are without intrinsic meaning. It makes no difference whether the content they refer to is trivial or of geopolitical urgency. This is because all traces that can be turned into data are apt to answer questions they were not intended nor imagined to be answered. Claude Shannon, to whom we owe the theory of the transmission of information, was the first to point out that the content of transmissions was irrelevant to the act of transmission. His definition of information counts only the number of bits required for the transmission of the system's state.

So what does big data do to time and our sense of time? It operates in the temporal regime of electronic data processing, which is not directly accessible to our temporal perception, biologically limited as it is. But being part of global electronic networks enables it to absorb all the traces we leave, provided they can be transformed into data. The traces are many. They include the minutest trivia of everyday life: the countless social interactions that connect us with our fellow human beings on a daily basis; the traces left by our reading and listening habits; our consumer behavior; the number of steps we have taken and our changes of location, plus whatever means of transport we have used. And so on and so forth.

What do we receive from big data in return? Targeted information, customized to our individual profile. What kind of information? Information about anything and nothing in particular. About things that are supposedly good for us and that we should therefore be doing. We even pay for this information while we supply the required data for free. What we receive in return is a multifaceted offer, a glut of options: purchase this book, that gadget, those designer shoes. This is where you want to go for your next holiday. Do this for the sake of your health, and if you engage in sports, do exactly that which will improve your performance.

Big data records the past indiscriminately. The more traces, the more data, and therefore, the better. Content does not matter. Questions concerning the *Why?*, the motivation, are replaced by questions of *What*. Only *What* questions generate data that are fit to be used over and over again—for other purposes, other questions, in other contexts. This transforms meaningless data into something that has meaning and significance. Predictive analytics is all about enabling predictions of our behavior in the present and in the immediate future based on our behavior in the past—in other words, on the traces we have left.

Big data homogenizes the past. Since all data are, in principle, fit to be used for purposes as yet unknown, the past is deprived of depth of focus. It becomes an abstract segment of time, a t_1, located on an equally abstract timeline. A discrete event becomes a data point, one among millions.

Connections established by algorithms make visible the relationship between patterns of behavior in the past and in the future. The information supplied to us by the commercial operators of the big-data networks contains a personalized promise for the future to be redeemed as soon as possible. Given this deluge of information, it is little wonder that news items, messages, instructions, and promises vie desperately for our attention all the time.

Attention, as has often been noted, has become an economically scarce resource. The big internet corporations in charge of the data networks are ingenious in turning this scarce resource into as much profit as possible.

As users, we find ourselves confronted with the question of whom or what we should lend—or sell—our attention to. This reawakens the longing for the now in a new, unexpected manner in the midst of everyday life saturated with information and communications technologies. For some, the solution lies in mindfulness, in focusing consciously on oneself, on breathing, on one's thoughts—or the absence of thought. Mindfulness is consciousness that is purposefully directed at the present, at the here and now; at calling for a break in the routine of living; at interrupting the familiar way in which time passes; at immersing oneself for a short time as a different way of experiencing time.

This is one more reason why music lets us enter into a different time frame. Music presupposes mindfulness. It is full of intervals and pauses. In some pieces by Johann Sebastian Bach, the interval is longer than the note that precedes it. Longing for the moment, however, is not to be satisfied only by engaging in mindfulness or listening to Bach sonatas. It is too deeply inscribed in our existence. Inexorably, the biological arrow of time points forward. Mindfulness would like to wrest moments from it that are not subservient to its time regime, notwithstanding the certainty that time's arrow will ultimately overtake all moments.

As so often in the history of humanity, technology intervenes. It alters our relationship not only with the environment and our fellow human beings but also with ourselves. It alters our experience of *eigenzeit*. The endless multiplication of the amount of electronic data and the increasing density of the networks create a new, technology-based set of options to satisfy the longing for the moment: medial *eigenzeit*.

The technological building blocks reside in the countless apps programmed by the algorithms that underpin the connections from which medial *eigenzeit* flows. It is both standardized and personalized. The offers it makes are filtered and selected—customized—for you, your lifestyle, and your preferences. Medial *eigenzeit* is accessible from everywhere via an iPhone or smartphone. It is ubiquitous, and it manages both to be available individually to each of us and to connect us to others. This is what makes it so attractive.

It seems that what I wrote in the last chapter of my book almost thirty years ago has come true in a completely unexpected manner. The longing for the moment, the longing to have time at one's disposal to be used at one's discretion, finds its technological fulfillment. Medial *eigenzeit* is available at any time. A digital switch, a key with an on/off option, is all that is needed. We are in full control of the moment, knowing that the next one can be switched on at any time.

Immersing oneself in medial *eigenzeit* grants time to indulge in exchange with others. Rather than the conscious void that is being sought in the practice of mindfulness, time is full of information—about the world, about friends, even if these are "friends" only in the social-media sense; information about oneself and those parts of it that we want to share with others. It gives us a sense that we are incessantly engaged in communication with the entire world.

Medial *eigenzeit* is time one has to oneself in order to spend it with those who are absent. It spells the end of boredom, that feeling that time is dragging on, that nothing of interest is even remotely in sight. For centuries, people have sought refuge from boredom in distraction. Now distraction, entertainment, and the conviction that all the latest developments are within easy reach are built into those small devices that channel medial *eigenzeit* into one's fingertips.

What does the disappearance of boredom mean for the development of children? Who does not remember from their childhood those episodes of indefinably empty stretches of time that seemed unending? They generated a restlessness that became the breeding ground for ideas and for determined efforts to engage with something new. Can one really grow up—in any meaningful sense of the phrase—without ever having experienced bouts of boredom?

Sherry Turkle has conducted in-depth interviews with American youngsters and with managers of hi-tech companies. These interviews permit insights into a world where all that matters is instant communication on a broad front. The youngsters practice what Turkle has called “continuous partial attention.” The capacity for conversation and the command of body language are being lost. This new generation has problems listening to others and developing empathy. Turkle, however, does not yet consider conversation a lost cause. We should treat communicative behavior as we treat eating—to be conducted with a set of rules and in moderation.[9]

Rules have to be formulated by someone, and observing them requires self-discipline. Here, too, medial *eigenzeit* promises relief. It supplies us with direct information about our behavior in ways not previously available.

Countless apps now offer observation of the self in real time. Biomarkers—measurable indicators of vital signs and various bodily states—are compared with standard parameters and/or yesterday's measurements. One can enter performance goals for the self that one would like to be tomorrow. Medial *eigenzeit* can be further expanded with the help of new wearable electronics, textile garments with integrated electronic devices. The aim is to monitor around the clock the functions of the entire body.

Biomarkers are "selfies" taken by our vital functions. They make the bio-temporal fluctuations of our inner rhythms visible. This data, too, is passed on continuously to be combined with other data for ever-new purposes. Each individual is both connected in collective real time to countless others and simultaneously set apart as a unique human being. Uniqueness only makes sense if there is a frame of reference composed of the uniqueness of the others. The self feeds its data into the collective in order to be confirmed as a self by the collective. This is yet another self-referential loop inherent to and enabled by medial *eigenzeit*.

It is also one of the reasons why medial *eigenzeit* fits seamlessly into the "sharing economy" and merges easily with what has been termed "open innovation." This creates an all-round win-win situation: the big internet corporations are tremendously profitable; small start-ups are encouraged to put their latest apps on the market and not to be deterred by the probability of failure. Even users stand to benefit.

One of the benefits that should not be underestimated is the consolation medial *eigenzeit* extends to those in need. Whenever you are overcome by the anxiety of continued acceleration, of things spinning out of control, and by the realization that you may not be able to achieve the goals you have set for yourself, you can make use of the communicative gift of the moment: you are not alone in being plagued by these anxieties.

But even though medial *eigenzeit* continues to proliferate in our lives in countless variations, it is limited in duration. Even if it devours the greater part of the day, there is more to everyday life than medial *eigenzeit.* Unexpected things happen. The vicissitudes of life are unconcerned with temporal arrangements. And time's arrow points steadily forward, unmoved by anything else.

The moments we spend with *eigenzeit,* whether in its virtual or its real variant, are part of life time. *Eigenzeit* and life time are bound together in a mutual entanglement, marked by a productive tension that defies resolution. Given the finitude of one's life time with its unknown expiration date, the yearning for the present moment and the wish to have time for oneself expresses our deeply conflicted longing for the duration of human existence.

This tension is the product of life time's harsh and undeniable facts and the wish to preserve that which makes life worth living. *Eigenzeit* is set to mediate—between growth and decay, between disease and death on the one hand and *joie de vivre* on the other. It grants us temporary unity with the self while always remaining precarious. Such unity must always be wrested from life time and requires constant reintegration.

Only at first sight does this seem not to apply to medial *eigenzeit*, given its tendency to shy away from tensed states with the help of "instant" communication, often in vain. The self is constantly being reconstituted medially by social interaction. Its identity consists of a multiform, variable entity, but there is no escape from the limitation imposed by a finite life time. Yet the constantly recurring constitution of the medial self opens up a new level of self-reflection. The self is in a position to observe itself and to intervene. It can share with others how the interactions with them change the self's embedding in its life time.

An overdose of medial *eigenzeit* leads into the blind alley of egotism, where people can no longer reach out beyond themselves and lose the capacity for empathy. "Friendships" that hardly scratch the surface cannot conceal the fact that the self may be hollowed out. Life time loses its resilience. Reflecting on the tension between medial *eigenzeit* and life time may, however, also lead to the emergence of new forms of solidarity. It can make people connect with others, provided their self is open to others and ready for forms of cooperation that cannot yet be envisaged.

However imaginative and creative the individual strategies may be that link *eigenzeit* and life time, they prove insufficient when confronted by the time scale of institutions or societies. It suffices to point to the acceleration of the human destruction of the natural environment to realize how difficult it is to fit the human lifespan into larger time scales.

How well equipped are current institutions to oppose the growing time pressure and the expectation of people's constant accessibility? What is their contribution to making the tension between life time and *eigenzeit* productive? Institutions are time transformers. In principle, they command all the resources and the legitimacy required to coordinate, balance, and adapt the temporal horizons and regimes of all parties concerned. What is lacking today, however, is institutional spaces, free to experiment with new ideas. Institutions lack the courage to experiment with different temporalities.

No time is set aside by institutions for reflection on these dilemmas or for institutional self-reflection. The institutions we have today were created as frameworks of reference and as means of regulation for the problems of the past. Their capacity for providing the kind of "temporal governance" required today is woefully underdeveloped. First, the long-term perspective is missing. Second, they lack what it would take to counter the neoliberal pressure for improvements in efficiency. This applies to many of our contemporary problems, from climate change to the threat that automation poses to the labor market.

In 1930 the renowned economist John Maynard Keynes gave a lecture at the students' club of the University of Madrid in which he shared his thoughts on the economic future awaiting the generation of his grandchildren. Most of these predictions, made for a timespan of one hundred years regarding economic growth and affluence, have since come true. One prediction, however, baffles us today: Keynes prognosticated that the average working day in 2030 would last only three hours.

Meanwhile, automation is progressing rapidly. Machines are learning to do what human beings do and are getting better at it all the time. It hardly needs saying that they take infinitely less time and soon will be encroaching on middle-class jobs. It is not 2030 yet, but the urgent questions concerning the societal reallocation of work and the separation of work and income remain unresolved.

In retrospect, it seems that the politics of time, understood here as the search for an alternative distribution of work and leisure, has lost its relevance. The considerations I put forward in *Eigenzeit* were based on a broad political articulation. It included the demands of workers for greater time sovereignty, sociopolitical arguments in favor of a work-life balance, and a more just distribution of paid and unpaid work between women and men.

Today the demand for more jobs, regardless of their quality, dominates politics. Consumer leisure time is on the increase in proportion to the growing online availability of consumer goods and services. Working hours have become more variable and more flexible. Largely unpaid time in care work is increasingly exposed to pressure for rationalization, and relief is being promised by ever-newer time-saving technologies. Was the political debate of thirty years ago merely a u-chronia, a "no-time"? And what changes have reshaped the relationship between power and time?

A subtle distinction enjoyed in the past by those in power was the license to keep others waiting. Today, power manifests itself in the license to make others work hard to survive in the highly competitive atmosphere of technological acceleration and in the creation of new markets. Power is in the hands of those who succeed in transforming their temporal-economic competitive advantage into protected monopolies and oligopolies. Power manifests itself by skillfully expanding successful products and designs to new areas. One example is Apple transforming the mobile phone into an all-purpose computer. The corporation succeeded in creating a completely new entertainment system and to invent a new kind of shopping mall. Even though we are continuously told that "disruptive" technologies will wipe out all competitors, the tenacity of those in power to cling to it is remarkable.

One of the central theses of *Eigenzeit* concerned the dissolution of the demarcation line between "present" and "future" as categories. A present predicated on—even obsessed by—innovation in science and technology and by innovation for its own sake appropriates the future. The present is turned into an arena for the deployment of knowledge, and a wide range of activities are put into place to filter and select the countless development options offered by the overwhelming potential of the technosciences. To safeguard evolutionary selection, a radical reduction must take place that pares down these options to a few. This also results in an alteration of the quality of time.

The time paths of processes leading into the future thus become more stable and increase in density. Temporal dependencies are constituted. "The extended present has chosen the future and not vice versa."[10]

With the end of modernity's belief in linear progress, the category of the "future" as a clearly envisaged goal has lost much of its attraction. Today the future seems more fragile than ever before. The loss of the great historical horizon and the predominance of the West have fragmented it even more. The categories of past, present, and future do not have universal validity. Today they are undergoing another shift. The future is appropriated by the extended present. The intensity with which the future is evoked and the countless predictions that are being made are but steps in the process of its appropriation.

William Gibson, a well-known sci-fi author, once noted laconically: "The future has arrived—it's just not evenly distributed."[11] The future does not disappear, but its status has changed. Moving it nearer to the present has made its inherent uncertainties more clearly visible. Temporal complexity is generated through the unpredictable combination of the different components of the system, the different time scales involved, and the different speeds at which they operate. Its dynamic is difficult to predict. Again, we are being made acutely aware of how limited our capacity is to foresee—let alone control—the unintended consequences of human action.

An extended present has particular repercussions on the temporal experience of the past. Some of the resources provided by that experience are lost, while new ones are gained. The past shrinks—it is leveled and appended to the present, a process largely effected by science and technology. The digitalization of past events opens up spectacular didactic opportunities and a new approach to these events. Visitors of Pompeii, Stonehenge, Laas Geel, and other sites of antiquity can log in digitally at these sites, allowing them to follow the day-to-day lives of the sites' former inhabitants and to take part in their (reconstructed) rituals without ever actually setting foot there.

The more the form of the representation and of its interpretation assimilates what is familiar and therefore contemporary, the more the distinctions vanish. The past stops being a "foreign country" in which "they do things differently" (as has famously been said), with alien structures and alien sensibilities. One of the most valuable resources of the past is in danger of being lost here: the knowledge that it could have been different.

This is compensated by gains. The more our own past, shaped by human history, is made part of the present, the more the distant past is made visible with scientific precision. Not only can we look further back into the past, we also see more.

The knowledge generated by paleogenomics and the methods at its disposal today are completely rewriting the evolutionary history of our species. New relationships between us and the Neanderthals and Denisovans have been discovered. We learn that parts of today's Amazon basin were probably first settled by populations that had made their way there from the Australian archipelago at a time when the distances were much shorter and the sea level much lower. We thus become aware of long periods of time, encompassing thousands of years, as though they were yesterday.

To an even greater extent, this is true of space, which has also become part of the extended present. The impressive pictures of Pluto relayed back to earth in July 2015 familiarize us with a celestial body in the far reaches of the solar system as if it were as close to us as the moon.

Space is increasingly emerging as the vanishing point where the past of the entire universe encounters the immediate future of humanity in an extended present. Behind us are the futuristic visions of the 1950s, the flying cars and other fanciful prognoses. Instead we have driverless cars and drones, which will proliferate as soon as the markets are ready. Robotics has made huge advances. The progress of artificial intelligence has begun to cause concerns even for the pioneers intimately involved in its development. Some assume that these rapid scientific/technological developments will result in the total loss of the future.

This is a view I definitely do not share. We live in an extended present that has appropriated many of the possibilities that we used to think of as located in the distant future. It brings us closer to the time scales of the universe and of evolution that exist independently of human activities but are made accessible through human activities. The epoch of the Anthropocene, marked by the undeniable impact of human intervention on the natural environment, forces us into a cognitive and emotional confrontation we can no longer escape.

Arjun Appadurai has called the future a "cultural fact." He reminds us of the millions of people in India who, for the first time ever, are able to articulate their wishes and longings. Finally, they succeeded in having the "capacity to aspire."[12] Perhaps this applies to us in the saturated West as well.

Our ideas of what a "good" society might look like are in danger of becoming stunted by the deluge of new apps, gadgets, and toys that indulge our predilection for technology. Do we have consensus answers to the question raised in antiquity about *eudaemonia*, the good life we are striving to achieve not only for ourselves but for the societies we live in?

Science and technology release their tremendous potential only in co-evolution with society. The technosphere, this assemblage of technologies and their economic, societal, and cultural preconditions and dispositions, unfolds a power that transforms reality. It reduces the distance between present and future in an unprecedented way. The reason we will not lose the future is simple: there is no predetermined future, only a future that is as radically open as it is inherently uncertain.

I have reached the end of revisiting *Eigenzeit.* It leads me to a new encounter—with the Cunning of Uncertainty.[13]

Translated by Otmar Binder

Endnotes

1 Helga Nowotny, "Time Structuring and Time Measurement: On the Interrelation between Timekeepers and Social Time," in *The Study of Time*, vol. 2 (Berlin: Springer Verlag, 1975), 325–342.

2 Helga Nowotny, *Time: The Modern and Postmodern Experience* (Cambridge: Polity, 1994, first published 1989), 11.

3 Hartmut Rosa, *Beschleunigung und Entfremdung – Entwurf einer kritischen Theorie spätmoderner Zeitlichkeit* (Berlin: Suhrkamp, 2013).

4 Judy Wajcman, *Pressed for Time: The Acceleration of Life in Digital Capitalism* (Chicago: University of Chicago Press, 2015).

5 Helga Nowotny, "Times of Complexity," in *Time, Order, Chaos* (The Study of Time IX), ed. J. T. Fraser, Marlene P. Soulsby, and Alexander J. Argyros (Madison, CT: International Universities Press, 1998), 91–146.

6 Bernd Scherer, "Wider die Herrschaft des Augenblicks," in *Die Zeit der Algorithmen* (100 Jahre Gegenwart), ed. Bernd Scherer (Berlin: Matthes & Seitz, 2016).

7 Jacques Attali, *Une brève histoire de l'avenir* (Paris: Fayard, 2006–2007).

8 Helga Nowotny, *Time*, 10.

9 Sherry Turkel, *Reclaiming Conversation: The Power to Talk in the Digital Age* (New York: Penguin Press, 2015).

10 Helga Nowotny, *Time*, 52.

11 William Gibson, *Economist*, December 4, 2003.

12 Arjun Appadurai, "The Capacity to Aspire: Culture and the Terms of Recognition," in *Culture and Public Action*, ed. Vijayendra Rao

and Michael Walton (Palo Alto, CA: Stanford University Press, 2004), 59–84.

13 Helga Nowotny, *The Cunning of Uncertainty* (Cambridge: Polity, 2015).

For Product Safety Concerns and Information please contact our EU representative GPSR@taylorandfrancis.com Taylor & Francis Verlag GmbH, Kaufingerstraße 24, 80331 München, Germany

Batch number: 10726001

Printed by Printforce, the Netherlands